Grade 1 · Unit 1

All About Plants

FRONT COVER: (t)showcake/iStock/Getty Images, (tr) Butterfly Hunter/Shutterstock, (b)Vladvm/Shutterstock; **BACK COVER:** showcake/iStock/Getty Images

mheducation.com/prek-12

STEM McGraw-Hill is committed to providing instructional materials in Science, Technology, Engineering, and Mathematics (STEM) that give all students a solid foundation, one that prepares them for college and careers in the 21st century.

Send all inquiries to:
McGraw-Hill Education
8787 Orion Place
Columbus, OH 43240

IBSN: 978-0-07-699613-1
MHID: 0-07-699613-1

Printed in the United States of America.

6 7 8 9 10 11 LMN 26 25 24 23 22 21

Table of Contents
Unit 1: All About Plants

Plant Structures and Functions

Plant Parents and Their Offspring

Plant Structures and Functions

DISCOVER
THE PHENOMENON

How does this plant stay upright?

Plant Parts

Check out *Plant Parts* to see the phenomenon in action.

Talk About It

Look at the photo.

Explore the digital activity.

What do you observe?

What questions do you have?

In some plants, the parts underground are larger than the parts we can see above the soil!

What Does a Landscape Architect Do?

GO ONLINE Learn about a Landscape Architect.

Landscape architects design outdoor spaces. Parks and playgrounds are outdoor spaces. Landscape architects must know how different plants grow.

1. What do landscape architects need to know about plants?

2. Think about an outdoor space where you live. Draw a picture of it. Name the outdoor space.

3. **ENVIRONMENTAL Connection** Explain something a landscape architect can do to make sure outdoor spaces stay safe and healthy.

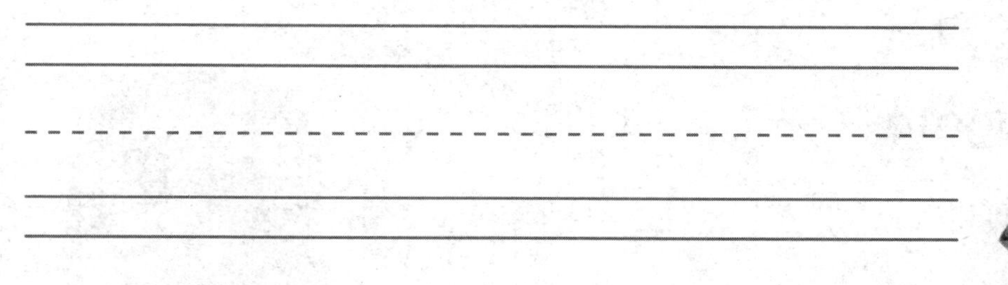

Tell a partner your ideas.

KAYLA
Landscape Architect

Plant Structures and Functions

Here are some words you will learn.

flower 

fruit 

leaf 

root 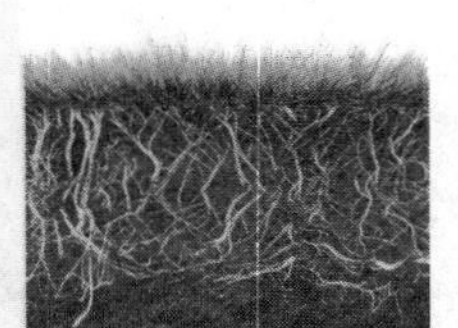

seed 

stem

Words to Know

data investigate

model pattern

You will learn these words during the lessons.

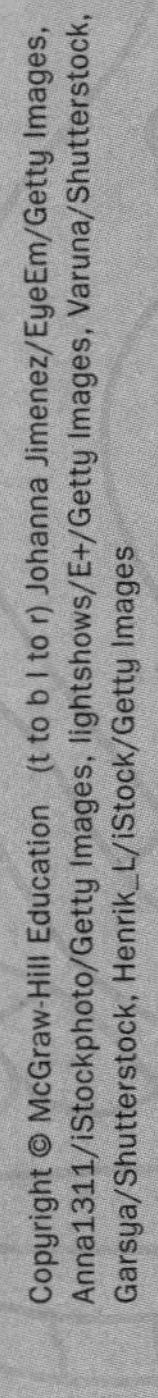

LESSON 1 LAUNCH

Parts of a Plant

Circle the parts of a plant that help it live and grow.

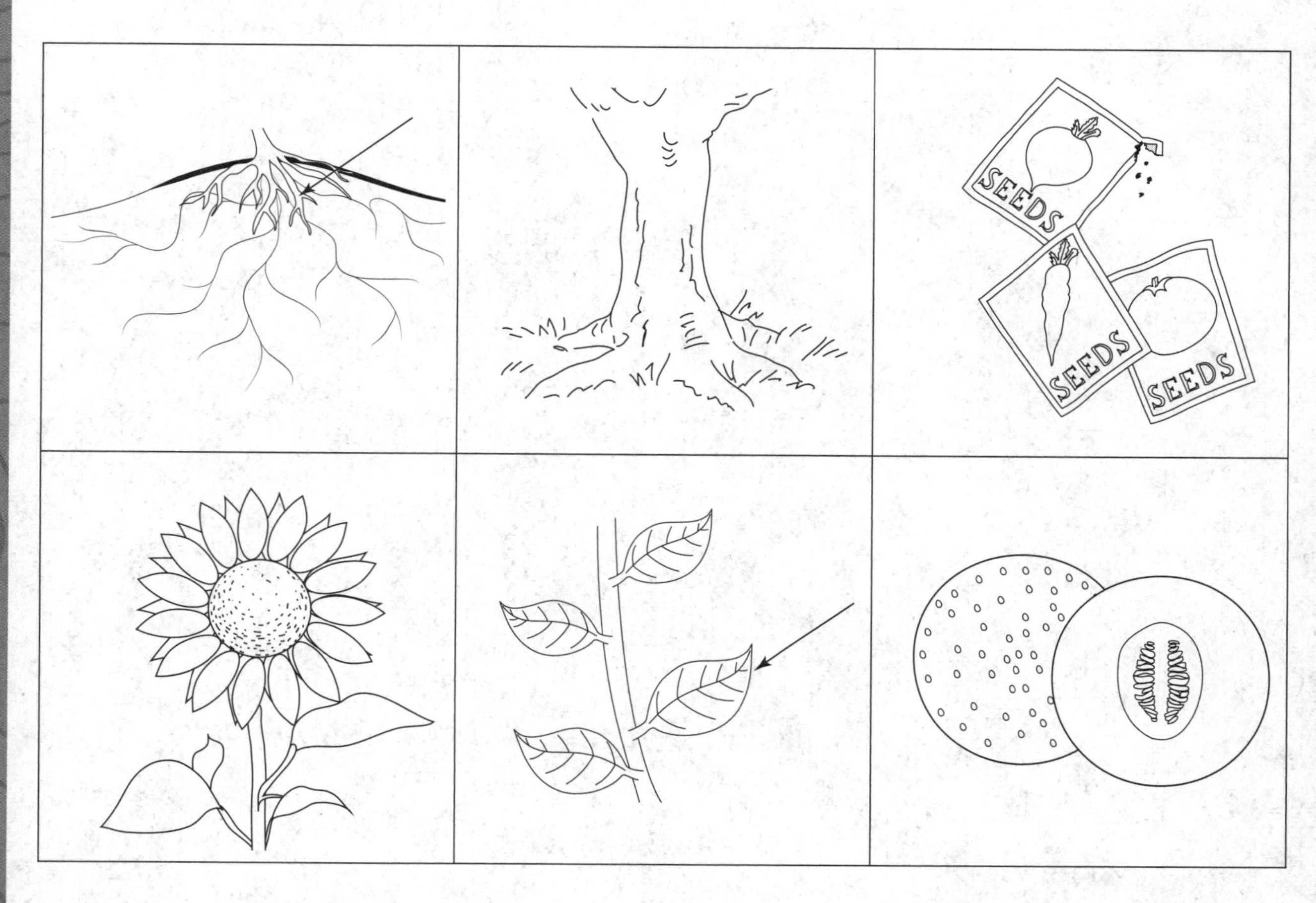

Explain your thinking.

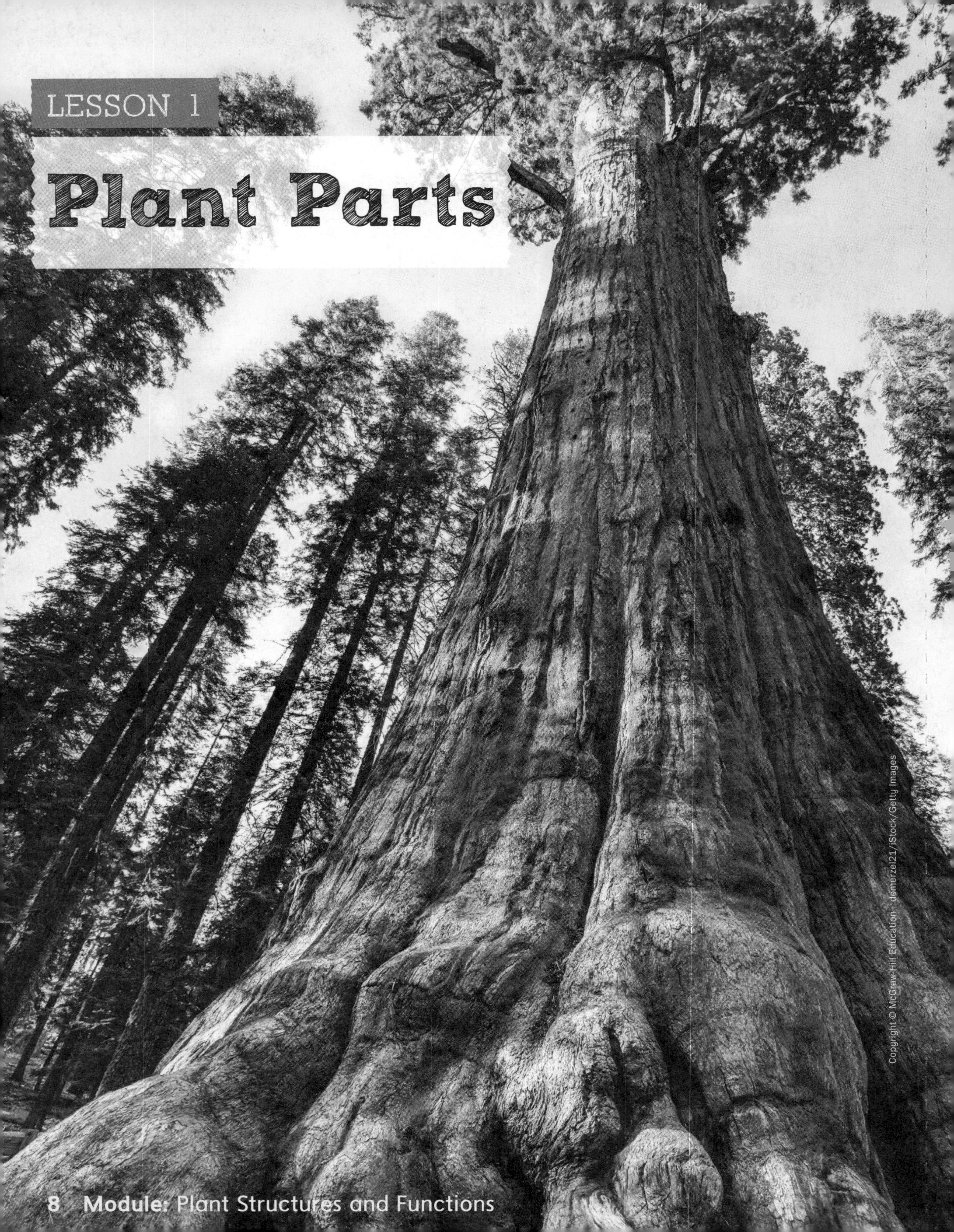

LESSON 1

Plant Parts

DISCOVER
THE PHENOMENON

How is this sequoia tree different from other plants?

Check out *Sequoia National Park* to see the phenomenon in action.

Look at the photo. Explore the digital activity. What parts of a tree do you know? What do you observe?

Did You Know?

This tree is very tall. It is as tall as a building with twenty-five floors.

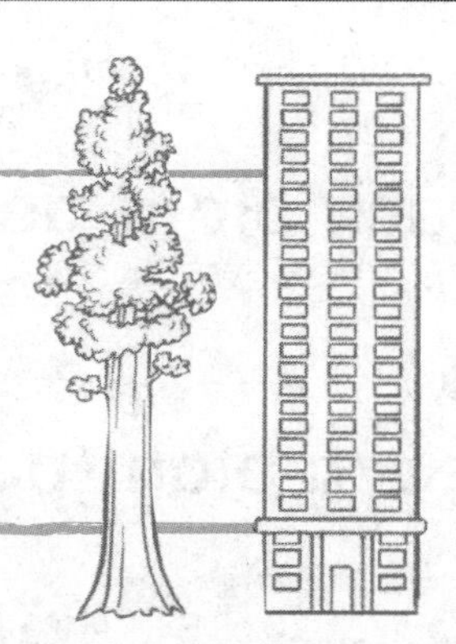

INQUIRY ACTIVITY

Hands On

Observe Plant Parts

You observed the parts of a tree. Observe parts of another plant.

Materials

 plant

 hand lens

 crayons

 flashlight

Make a Prediction How are parts of the plant different?

____

____

Investigate

BE CAREFUL Wear gloves.

1. Choose three different plant parts.
2. Draw a picture of each part in the table.
3. Use the hand lens. Look at each part. Observe the color and shape.
4. Use your hands. Carefully feel each plant part.
5. Use a flashlight. Shine light on each plant part.

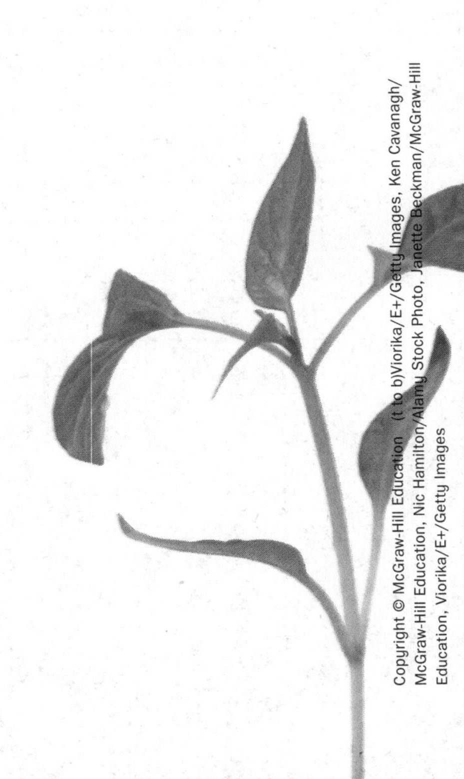

6. **Record Data** Write what you observe.

Plant Part	Observations

Talk About It

Why do you think the light only shines through some plant parts? Tell a partner.

INQUIRY ACTIVITY

Read *Plant Parts Around the World.*

7. Do any of Marco's observations about plant parts match what you observed in this activity? Explain using data.

8. READING Connection Where does this story take place? How do you know?

Talk About It

Which part do you think helps the plant stand tall? Tell a partner.

Make Your Claim

Do all plants have the same parts?

Circle the claim you agree with.

Claim
Plants always have the same parts.

Claim
Plants can have different parts.

Add evidence from this lesson to support your claim.

Evidence
I chose this claim because:

____

____

____

Discuss your reasoning as a class.

Reasoning
 **Talk About It** How does your evidence support your claim?

Vocabulary

Listen for this word as you learn about plant parts:

structure

Plants Have Parts

Watch the video *What Are Some Parts of Plants?* to learn more.

Plants have parts to get them what they need. Parts of the plants are also called structures. A **structure** is a part of something. Plants have many structures, or parts.

1. Draw a picture of a structure you know.

Explore *Parts of Plants* to see more plant structures.

Listen to *Comparing Plant Parts*.

2. Draw a line connecting each plant structure with its name.

flower: white and pink with many petals

stem: strong and tall, cylinder shape

leaf: green, feels waxy, curved

root: brown, looks like hair, in the soil

3. Why are some plant structures hard to see?

Inspect

Read the text.
Underline the text that tells about plant structures.

Find Evidence

Reread Highlight text that explains one way you can tell two plants apart.

Notes

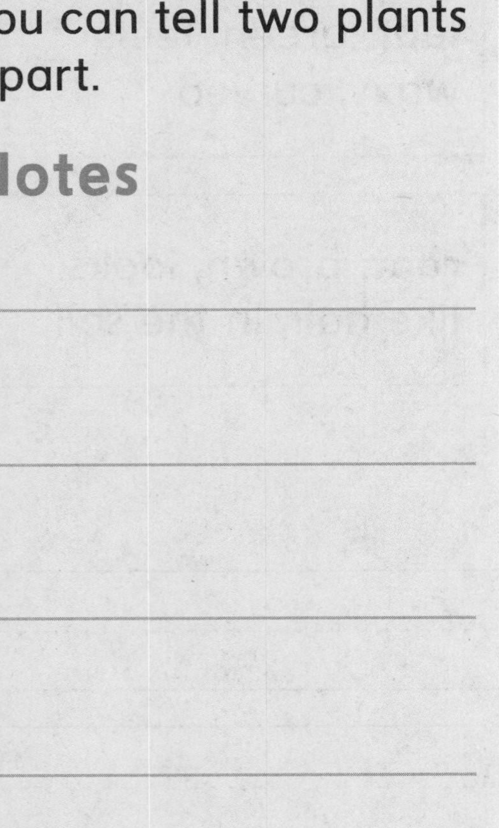

Plant Structures

Many plants have the same structures. A structure is a part of something. Leaves, stems, roots, and flowers are plant structures. Not all plants have the same structures. Plants can have different structures based on where they are found and what they need to live. A structure may also look different in some plants.

Make Connections

Talk About It

Draw a picture of a plant on a separate sheet of paper. Compare your plant to the one your partner drew.

Notes

You can use differences in structures to tell one plant from another. Some plants have flowers. The flowers can look different. You can use flowers to tell plants apart.

Look at the photos. How are these plant structures different? Explain.

INQUIRY ACTIVITY

Hands On

Plant Structures

Compare a daisy and an onion.

Make a Prediction Which structures do a daisy and an onion both have?

Materials

hand lens

onion

daisy

cubes

Investigate

BE CAREFUL Wear gloves.

1. Use the hand lens.
   Look at the structures of the onion.

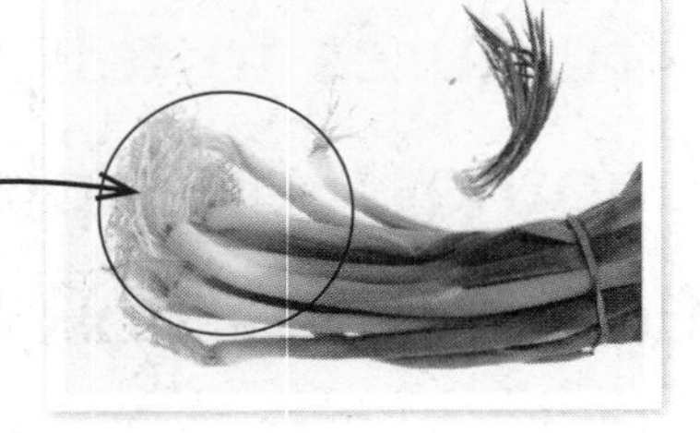

2. Use the hand lens.
   Look at the structures of the daisy.

This is an onion flower.

3. **Record Data** Put a ✔ in the box if the plant has the part.

	Root	Stem	Leaf	Flower
Daisy				
Onion				

Communicate

4. Does what you learned match your prediction? Explain.

____

____

MATH Connection Use cubes to compare the length of the daisy and the onion. Tell a partner what you find.

Trees Are Plants

Not all plants look the same. A tree is a plant. Trees have many structures. They have roots, leaves, and stems. The main stem of a tree is called a trunk.

A trunk helps hold up the tree. A trunk is usually covered with bark. The bark is a hard layer that keeps the tree safe. Trunks can be tall or they can be short. Trunks connect the top of the plant with the bottom of the plant.

1. Circle the trunk of the tree in the photo.

2. Circle the plant structure that is most like a trunk.

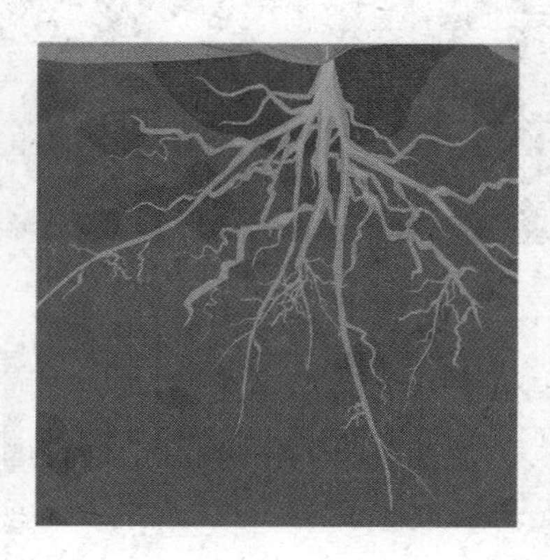

3. How are the parts of a tree different from the parts of a daisy?

What **structure** do a daisy and a tree both have? How are these **structures** similar?

LESSON 1

Review

EXPLAIN THE PHENOMENON

How is this sequoia tree different from other plants?

Summarize It

How are the structures of plants similar and different? Use your observations to explain.

REVISIT

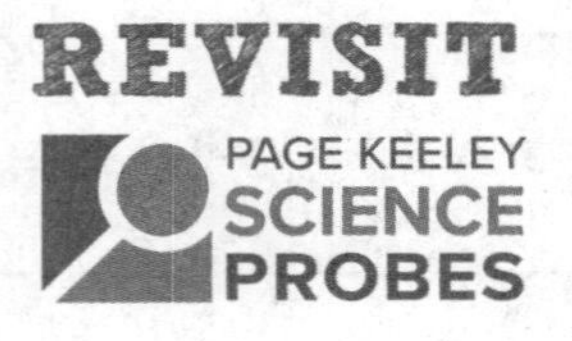

Look at the Page Keeley Science Probe on page 7. How has your thinking about plant structures changed?

Three-Dimensional Thinking

Answer these questions based on what you learned about plants.

1. Compare these plants. Circle the structures they both have.

2. Which statement is true about plant structures?

   a. Flowers can be used to tell plants apart.

   b. All plant parts look the same.

   c. All leaves are the same shape.

Extend It

OPEN INQUIRY

What questions do you still have about plant parts?

Write the question you want to investigate.

Plan and conduct an investigation to answer your question. Write the answer to your question.

LESSON 2 LAUNCH

Plant Part Functions

Which student has the best idea about the function of plant parts?

Liza: *I think the roots hold the plant in the ground.*

Jack: *I think the roots collect water.*

Deja: *I think the roots hold the plant in the ground and collect water.*

Explain your thinking.

LESSON 2

Functions of Plant Parts

DISCOVER
THE PHENOMENON

Where is the Sun in this photo?

Check out *Leaves Move* to see the phenomenon in action.

Look at the photo. Watch the video. Why do you think the leaves in the video move? What do you observe?

Did You Know?

Some plants will close or curl up at night!

INQUIRY ACTIVITY

Hands On

Plants and Light

You observed how the leaves of a plant follow the light across the sky. Investigate what other plants do.

Make a Prediction What will happen to the leaves of a plant as the Sun moves?

____

____

Materials

 plant

 crayons

Investigate

BE CAREFUL Wear gloves.

1. Put a plant in a sunny place.
2. Observe the plant in the morning.
3. Observe the plant in the afternoon.
4. **Record Data** Use pictures and words to record your observations.

5. Observe the plant for three days.
Add details.

Plant in the morning	Plant in the afternoon

Communicate

6. Did what you observe match your prediction? Explain.

7. How do you think the plant will move on the fourth day? Explain your reasoning.

Talk About It

Why do you think the leaves of some plants follow the Sun? Tell a partner.

PERFORMANCE EXPECTATION

STEM Module Project
Engineering Challenge

Build a Solar-Powered Light Stand

A park wants solar-powered lights along a trail. The solar panel on each light needs energy from the Sun. The stand for the lights should look like the other plants in the park. Help Kayla design and build a light stand strong enough to hold up a solar panel.

Materials

____

____

____

____

____

____

____

Design Your Solution

- [ ] Make a list of the materials you will use.
- [ ] Think about the structure and function of plant parts.
- [ ] Draw a design. Base your design on a plant structure.
- [ ] Use materials to build and test your model.

Use what you have learned to design your light stand!

STEM Module Project
Engineering Challenge

Build Your Model

Draw different pictures of what your light stand might look like. Circle the best one to build.

Test Your Model

Explain how well your design worked.

Module Wrap-Up

REDISCOVER THE PHENOMENON

How does this plant stay upright? Draw a picture. Add labels.

Plant Parents and Their Offspring

DISCOVER
THE PHENOMENON

What happens when you blow on a dandelion?

Dandelion Seeds

GO ONLINE

Check out *Dandelion Seeds* to see the phenomenon in action.

Talk About It

Look at the photo.

Watch the video.

What do you observe?

What questions do you have?

Did You Know?

The part of the dandelion that floats away when you blow it is the seed!

STEM Connection

What Does a Farmer Do?

Farmers grow plants and raise animals. Some farmers grow crops. Crops are plants that are grown and harvested. Farmers can grow grains, fruits, or vegetables.

Some farmers raise livestock. Livestock are animals raised for human use. Animals like sheep, cows, and chickens are livestock.

PRIMARY SOURCE

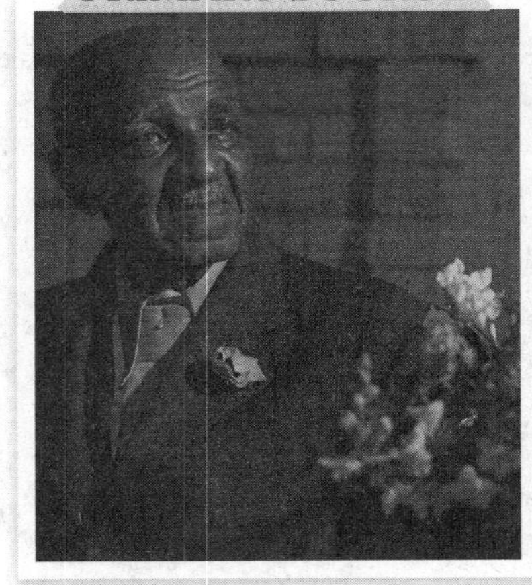

George Washington Carver was a botanist, inventor, and professor. He taught farmers how to grow crops that would keep the soil healthy.

1. Draw a picture of a something that comes from a farm.

The food we eat comes from farms.

2. **ENVIRONMENTAL Connection** Why is it important that farmers grow healthy plants and raise healthy livestock?

POPPY
Park Ranger

Plant Parents and Their Offspring

Here are some words you will learn.

need 

offspring 

parent 

seedling 

Words to Know

change

evidence

observe

LESSON 1 LAUNCH

Young Plants

Which friend has the best idea about young plants?

Joyce: *I think young plants look exactly like their parents.*

Melinda: *I think young plants look like their parents but can have some differences.*

Portia: *I think young plants look very different from their parents.*

Explain your thinking.

LESSON 1

Plants and Their Parents

DISCOVER
THE PHENOMENON

Are these plants the same?

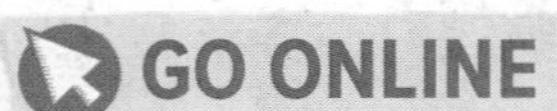

Check out *A Growing Plant* to see the phenomenon in action.

Look at the photo. Watch the video. How are young plants and adult plants similar? What do you observe?

Did You Know?
Almost all plants grow from tiny seeds!

INQUIRY ACTIVITY

Data Analysis

Compare an Adult Plant and a Young Plant

Plants change as they grow. Use the photos to observe how an oak tree changes.

Make a Claim How are a young and an adult oak tree alike and different?

____

____

Investigate

1. Look at the photos of the young oak tree and the adult oak trees.
2. **Record Data** Circle the parts that are the same.
3. Place an X on the parts that are different.

This is a young oak tree. It grew from a seed.

This is an adult oak tree.

Acorns grow on adult oak trees. The acorn is the fruit of the adult plant. There is a seed inside the acorn.

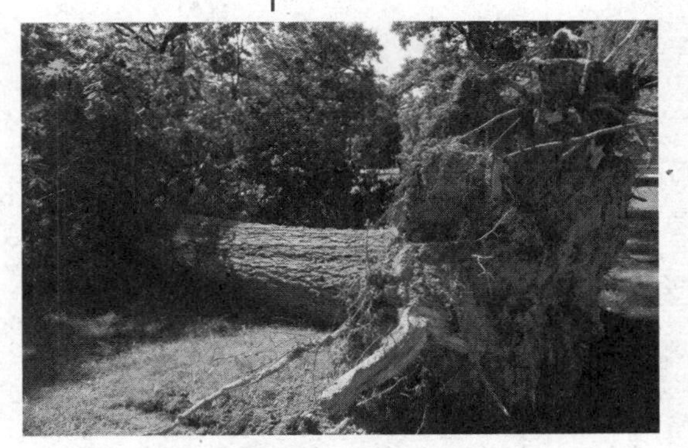

Strong wind blew this oak tree out of the ground. You can see the roots of the adult oak tree.

Communicate

4. How are the roots of the young and adult oak trees different?

5. What do you think would happen if the roots of the adult tree stayed the same size as the roots of the young tree?

6. Did your data match your claim? Explain.

How else do you think a young plant and adult plant are similar and different? Tell a partner.

Listen to *Perfect Acorn, Mighty Oak.*

Oak trees provide food and shelter to animals.

7. **WRITING Connection** Why does Gray plant an acorn and not a leaf?

8. How does your inquiry activity match what you read?

Vocabulary

Look and listen for these words as you learn about plants and their parents:

inherit offspring parent seedling

Seedlings Grow

A **seedling** is a young plant.
Seedlings grow into adult plants.

GO ONLINE

Watch the video *Plant Parents and Their Offspring* to learn about plants and their parents.

1. How does this plant change as it grows?

Listen to *Every Plant is Different*.

Young plants are similar to and different from their parent plants. You can observe these differences. Plants change as they grow.

2. How might a young plant be different from its parent? Make a list.

3. What do you think this sunflower looked like as a seedling? Draw a picture.

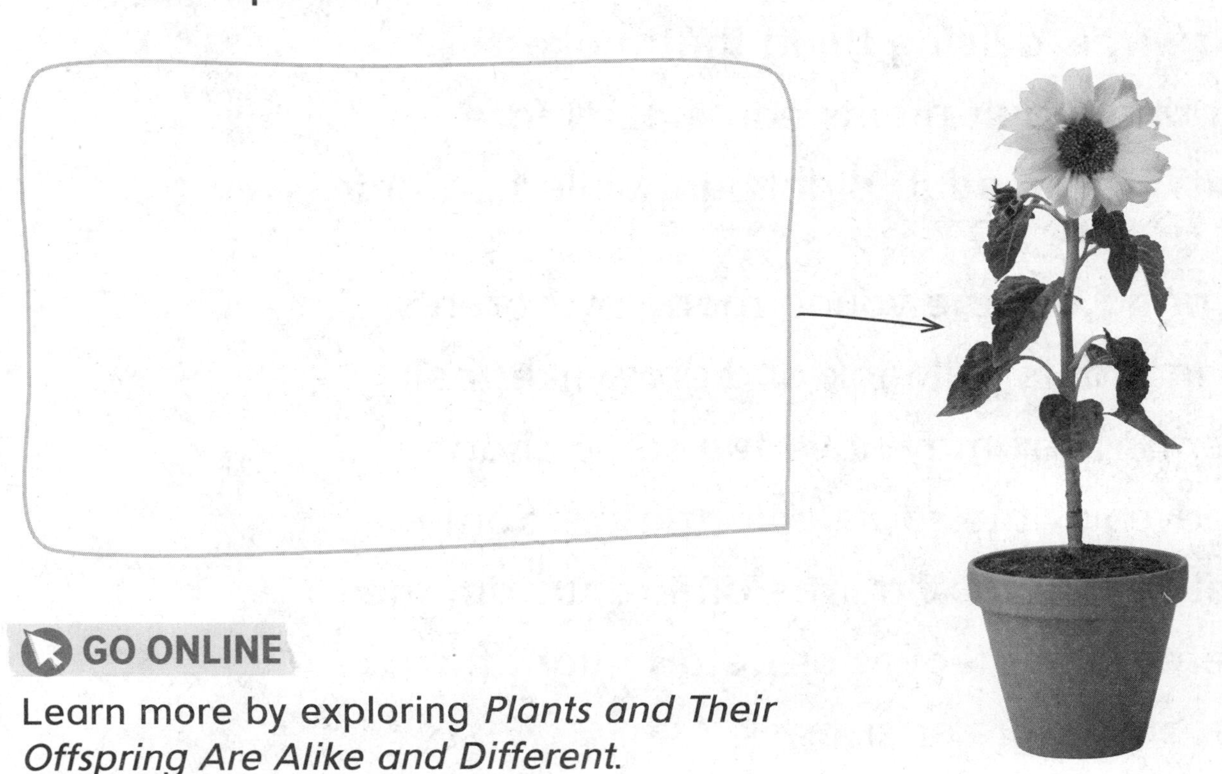

GO ONLINE

Learn more by exploring *Plants and Their Offspring Are Alike and Different*.

Offspring Are Like Their Parents

A **parent** is a living thing that makes offspring. Adult plants can make more plants. These adult plants are called parents.

Offspring are the young made by parents. Young plants will look and grow like their parents. They may have the same shape of leaves or the same color petals. Some plants might look a little different from their parents. Their leaves or petals might have different colors or sizes.

Inherit is when something is passed from the parent to its young. A plant can inherit its type of leaves or fruit. Offspring are like their parents. Offspring and their parents have the same structures. But sometimes plants from the same parents are not exactly the same. One plant may grow taller than another. Offspring are also different than their parents. Their flowers may be a different color.

Look at the photo of the tulips. ✓ how the tulips are different.

☐ color ☐ type of flower

Talk About It

What else do you think offspring inherit from their parents?

INQUIRY ACTIVITY

Simulation

Plants Grow and Change

Observe how green beans, carrots, corn, and peas grow into adult plants.

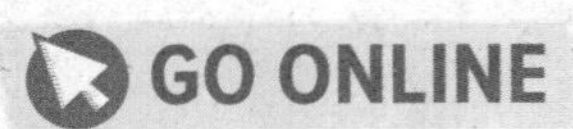

GO ONLINE

Use the simulation *Plants Grow and Change* to learn about how plants develop.

Make a Prediction How will the young plants change as they grow?

____

____

Investigate

1. Plant each vegetable in the garden.
2. Watch the plants grow.
3. Take photos as the plants grow.
4. **Record Data** Choose one plant and draw how it changes over time.

Plant	2 Weeks	4 Weeks	12 Weeks

Communicate

5. Did your observations match your prediction?

Think about the patterns you observed. Circle the structures that seedlings do not have. Explain your answer to a partner.

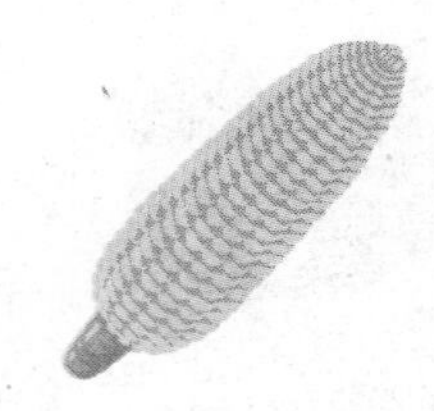

INQUIRY ACTIVITY

Hands On

Grow a Radish

Observe a radish as it grows.

Make a Prediction What will an adult radish plant look like?

Materials

plastic cup

potting soil

radish seeds

water

Investigate

BE CAREFUL Wear gloves.

1. Add soil to a cup. Plant two radish seeds in the cup. Cover with soil.
2. Water the plant when the soil is dry.
3. **Record Data** Record your observations for three weeks. Draw pictures.
4. After four weeks, dig up the radish. Record your observations.

Many people put this part of a radish in salads.

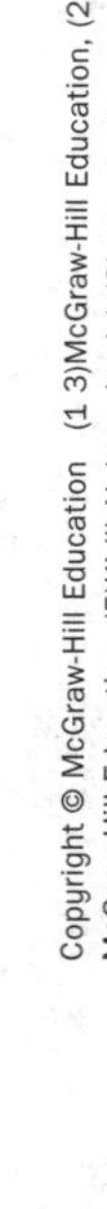

Week 1	Week 2
Week 3	**Week 4**

Communicate

5. Did your observations match your prediction? Explain.

LESSON 1

Review

EXPLAIN THE PHENOMENON | Are these plants the same?

Summarize It

How are adult plants and their offspring alike and different?

REVISIT

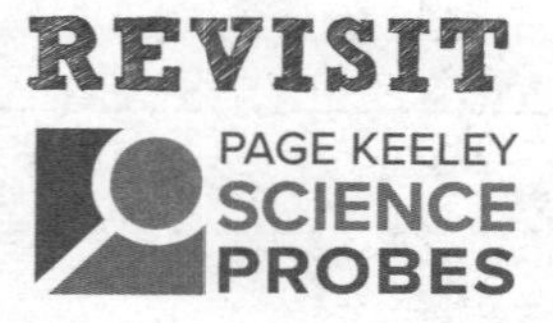

Look at the Page Keeley Science Probe on page 51. How has your thinking about young plants changed?

Three-Dimensional Thinking

Answer the questions based on what you learned about plants.

1. Which plant will this seedling grow into?

Circle the adult plant.

2. Offspring are exactly like their parents.

☐ true

☐ false

Extend It

OPEN INQUIRY

A friend wants to know how plants change as they grow. Choose a plant. Design an investigation to show how your plant will change as it grows. Write or draw your design below. Use labels.

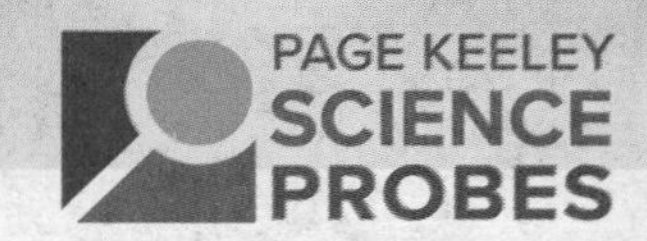

LESSON 2 LAUNCH

Plant Needs

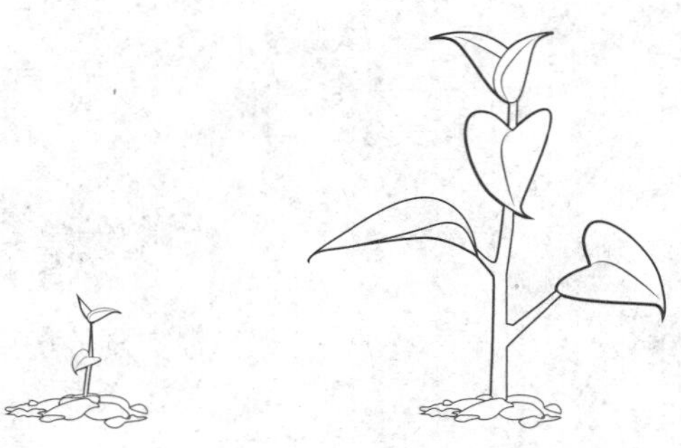

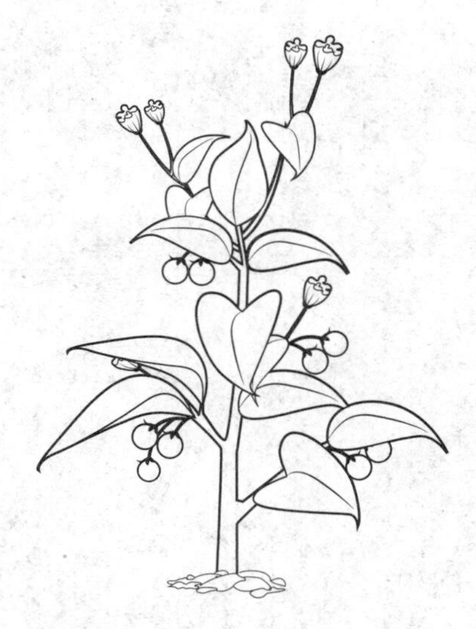

Which of the following are important for plants to live?

- [ ] Plants need water.
- [ ] Plants need a place to grow.
- [ ] Plants need sunlight and air.

Explain your thinking.

LESSON 2

Plant Survival

DISCOVER
THE PHENOMENON

What happened to this plant?

Check out *Changing Flower* to see the phenomenon in action.

Look at the photo. Watch the video.
How do plants stay alive?
What did you observe?

Did You Know?

The longest living plant in North America is a bristlecone pine. It is more than 5,000 years old!

INQUIRY ACTIVITY

Simulation

Plants and Shade

Plants cannot grow if they do not get what they need. Observe how two different plants grow in shade and sunlight.

GO ONLINE

Use *Plants and Shade* to see how plants grow.

Make a Prediction How will the amount of sunlight affect the growth of a plant?

____

____

Investigate

1. Plant three blue flowers in each part of the yard.
2. Observe what happens each year.
3. Repeat the simulation with the white flowers.

4. Draw a picture of the plant that grew best in the sunlight.

Communicate

5. Do all plants need the same amount of light? Use your observations to explain.

Talk About It

Do your observations match your prediction? Tell a partner.

Seeds

All plants need sunlight to grow. Some plants need more sunlight. Some plants need less sunlight. Seeds need things too. Seeds need water and a place to grow.

Listen to *A Little Seed's Journey*.

1. How did the little seed travel to a new place?

2. What do you think will happen to the little seed?

Make Your Claim

Do plants use their structures to live and grow?

Circle the claim you agree with.

Claim	Claim
Plants do not use their structures to live and grow.	Plants use their structures to live and grow.

Add evidence from this lesson to support the claim you chose.

Evidence

I chose this claim because:

____

____

Discuss your reasoning as a class.

Reasoning

Talk About It How does your evidence support your claim?

Vocabulary

Look and listen for these words as you learn about how plants stay alive.

need

pollen

survive

People Can Help Plants

PRIMARY SOURCE

People care for tomato plants in Imperial County, California in 1942.

To **survive** means to live and grow. Plants have needs. A **need** is something you must have in order to live. Water is a need because plants must have water to live and grow. Plants also need air and space to grow.

People can help plants grow. Sometimes people take leaves, stems, or branches off plants. This gives the plants more room to grow. This is called pruning. Pruning helps plants get what they need. Pruning helps the plant get more sunlight. It can make fruits grow larger.

GO ONLINE

Watch the video *How Plants Survive* to learn more about how plants grow.

Listen to *Making New Plants*.

Pollen is a powder that helps make new plants. Plants that have flowers or cones need pollen to make offspring. Pollen is very light. It can be spread by the wind or by animals. Some plants attract animals using color and smell.

1. How do animals and people help plants survive?

2. How do plants get animals to spread their pollen?

Inspect

Read the text.
Underline the things that help seeds travel.

Find Evidence

Reread Do all plants have the same type of seeds? Highlight the text that helps answer this question.

Notes

Traveling Seeds

Plants cannot grow in the same place as their parents. Plants have needs. Seedlings need their own air, water, and soil. Seeds do not have legs, so they get around in other ways. Some seeds travel in water. These seeds can usually float. Some seeds travel by wind. These seeds have structures that work like wings.

Make Connections

Talk About It

Look at the photo of seeds on the page. What kind of seeds do you think they are? How do you think they travel?

Some seeds are eaten by animals and dropped in new places. Other seeds can get caught on animals. These seeds have structures that help them stick to animal fur and even humans.

Draw a picture of a traveling seed.

Survival Structures

Plants have structures that help them survive, grow, and meet their needs.

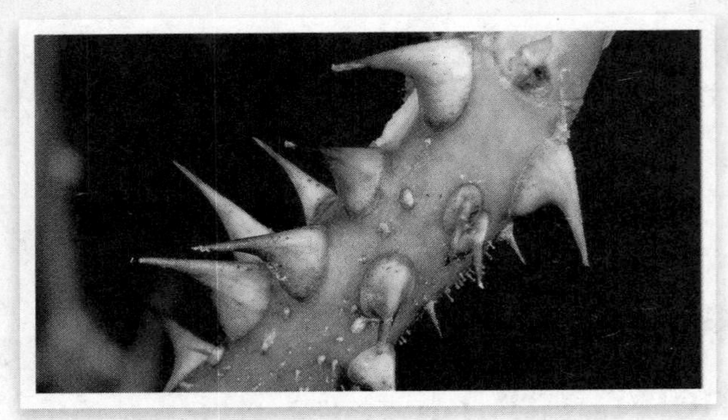

Some plants have thorns or prickles. These sharp structures stop animals from eating the plant.

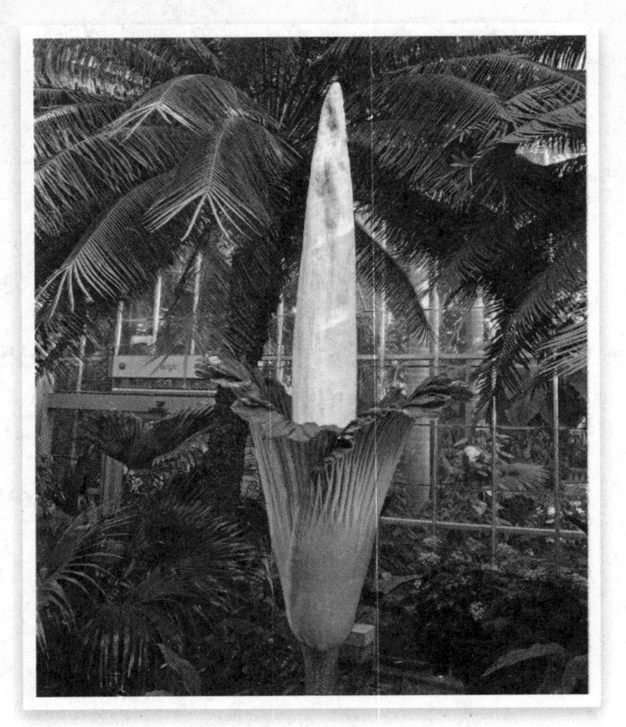

The corpse flower releases a strong smell like rotting meat. This smell attracts insects. The insects spread its pollen.

When the leaves of the dormilones plant are touched or shaken, they fold inward.

Chestnut seeds grow inside prickly burrs. This structure protects the growing seed.

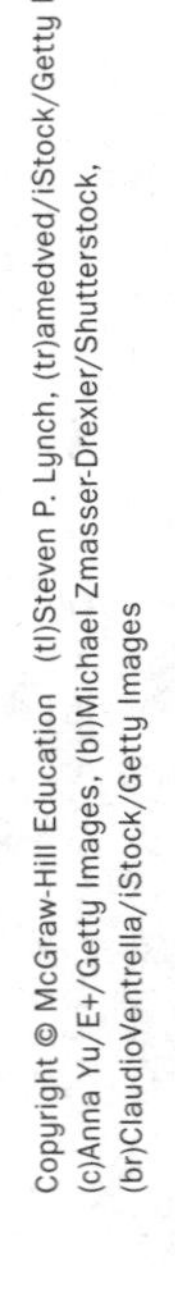

ENGINEERING Connection Copy the structures of a plant to solve a human problem. Explain the problem and draw a sketch. Use separate paper.

STEM Connection

What Does a Horticulturist Do?

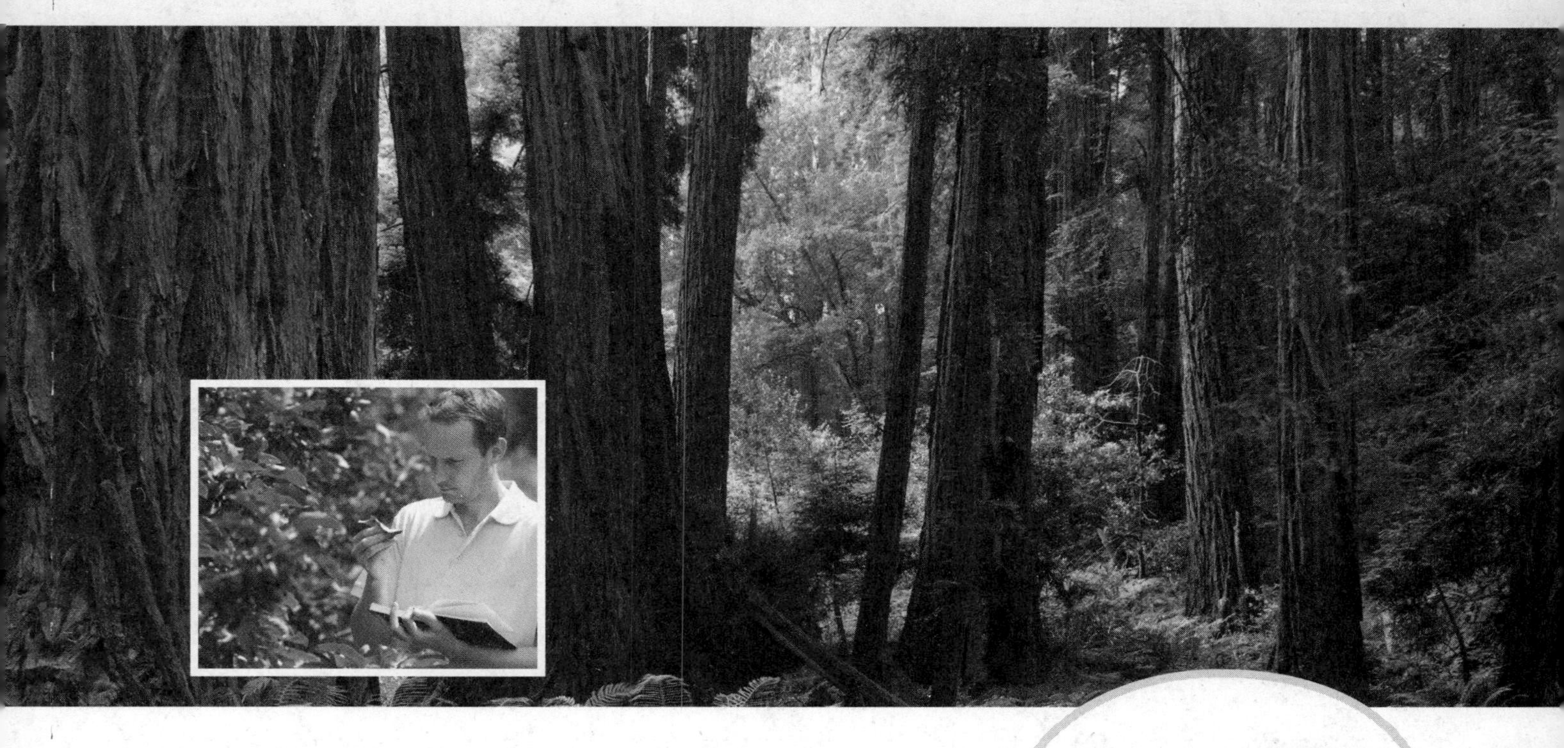

Horticulturists study plants. Horticulturists make sure plants and environments are healthy.

They help farmers grow better fruits and vegetables. Horticulturists work with farmers to protect the environment when they grow crops.

ENVIRONMENTAL Connection Why is it important to humans that Earth has healthy plants? Explain on a separate sheet of paper.

INQUIRY ACTIVITY

Research

OPEN INQUIRY

Plant Survival

Some plants have special ways to stay alive. Research how plants survive.

Ask a Question Choose a plant. Write a question you will research.

____

____

Investigate

1. Use books and online resources to learn about your plant.
2. Find the answer to your question.
3. **WRITING Connection** Explain how your plant survives on a separate sheet of paper.
4. Draw a picture of the plant you chose. Label the structures that help the plant survive.

Communicate

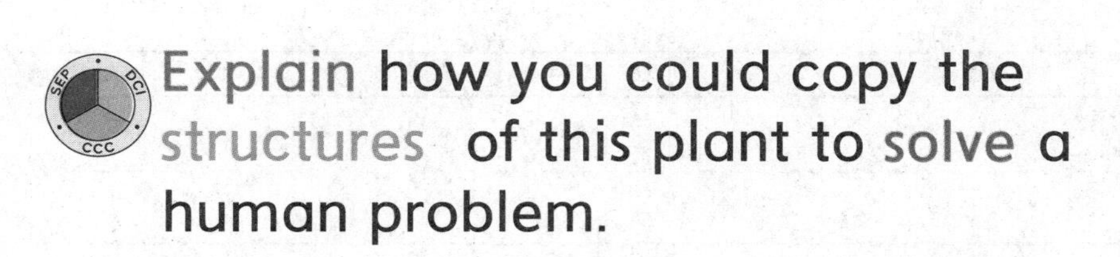

Explain how you could copy the structures of this plant to solve a human problem.

LESSON 2

Review

EXPLAIN THE PHENOMENON | What happened to this plant?

Summarize It

Explain what plants need to stay alive.

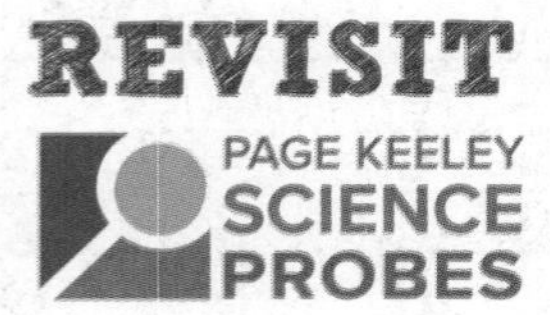

REVISIT Look at the Page Keeley Science Probe on page 69. How has your thinking about what plants need to survive changed?

Three-Dimensional Thinking

Use what you have learned to answer the questions about plant survival.

1. What would happen if plants did not make seeds? Place a ✓ in the box if the statement is true.

☐ If plants did not make seeds, they could not make offspring.

☐ Some animals might also go hungry without seeds.

2. How does the shape of a flower help the plant survive? Draw a model to help show what you know about structure and function.

Extend It

Use your plant research to write a newspaper headline. Draw a picture to show what you learned. Write a caption to explain your picture.

STEM Module Project
Engineering Challenge

Design a Seed That Travels

A farmer wants to design a way to spread seeds across a farm. Use what you have learned. Design a model seed to help the farmer.

Materials

____

____

____

____

____

____

Design Your Solution

1. List three ways seeds travel.

   ____

   ____

   ____

2. Circle how your seed will travel.
3. Draw your seed on the next page.
4. List the materials you will need.
5. Build and test your design.
6. MATH Connection Measure how far your seed travels.

Look back at the lesson to see how seeds move!

Build Your Model

Draw pictures of what your seed might look like. Circle the best one to build.

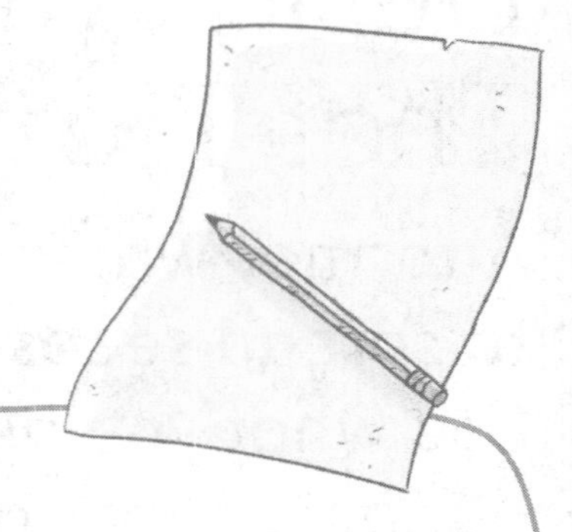

Test Your Model

Explain how well your design worked.

Module Wrap-Up

REVISIT
THE PHENOMENON

What happens when you blow on a dandelion? Draw a picture. Add labels.

Look at your project to help you answer the question.

Science Glossary

A

amphibian an animal that lives part of its life in water and part on land

B

behavior the way a person, animal, or thing acts or does something

bird an animal that has two legs, two wings, and feathers

C

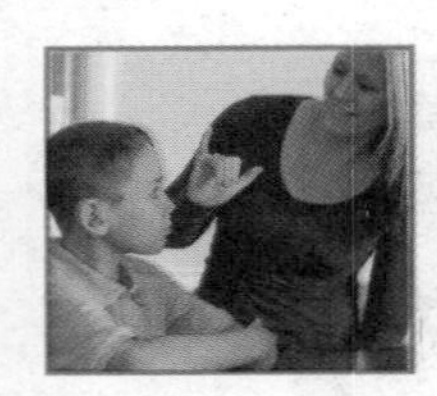

communicate to give information about something

E

Earth the planet on which we live

energy the ability to do work

F

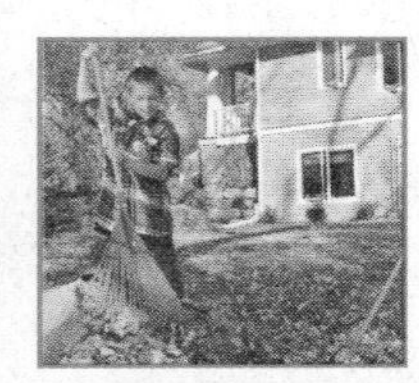

fall the season after summer

fish an animal that lives in the water and has gills and fins

flower plant part that makes seeds

fruit plant part that holds the seeds

function the purpose of something

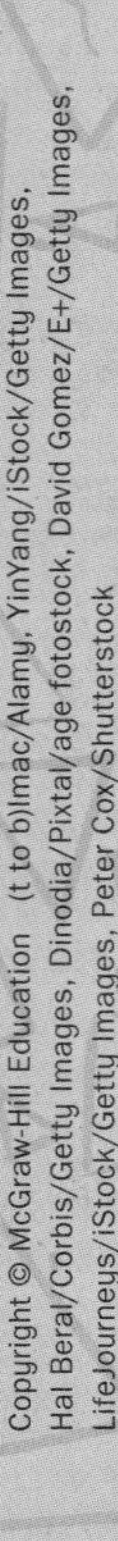

H

horizon the line where the earth and sky seem to meet

I

illuminate to light up

inherit when something is passed from the parent to its young

insect an animal with three body sections and six legs

L

leaf plant part that makes food from sunlight

learn to gain knowledge or skill

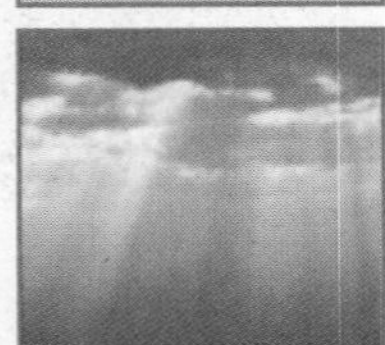

light a form of energy that lets you see

M

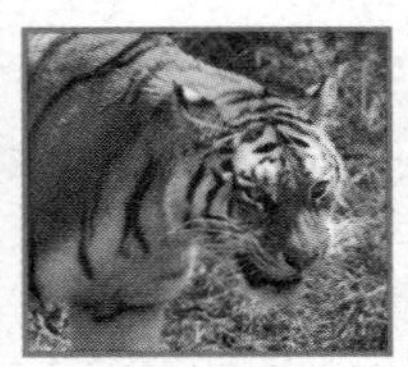

mammal an animal with hair or fur that takes care of and usually gives birth to live young

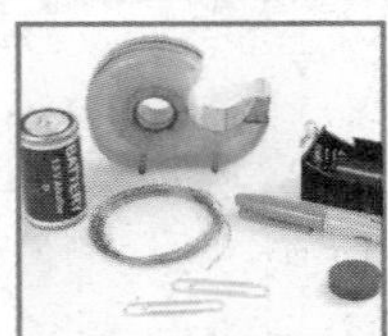

material what objects are made of

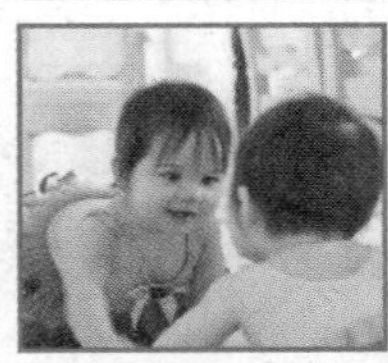

mirror a smooth surface that reflects what is in front of it

Moon a ball of rock that moves around Earth

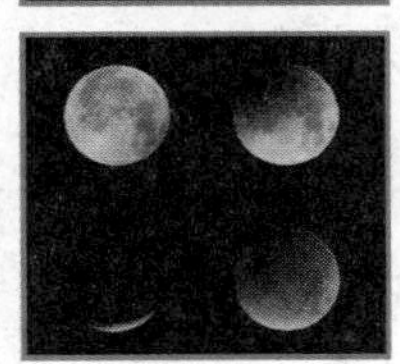

Moon phases the different Moon shapes we see each month

N

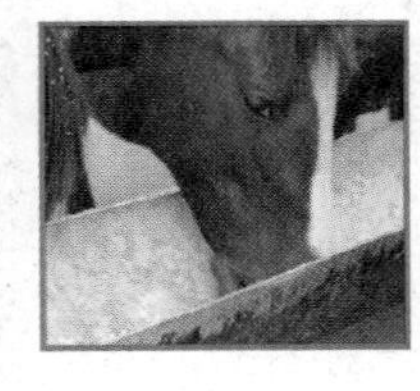

need something you must have in order to live

O

offspring young made by parents

opaque materials that do not let light pass through

P

parent a living thing that makes offspring

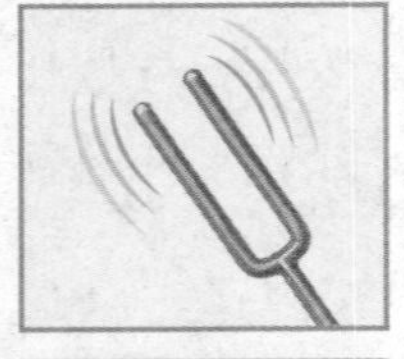

pitch how high or low a sound is

planet a very large object that moves around the Sun

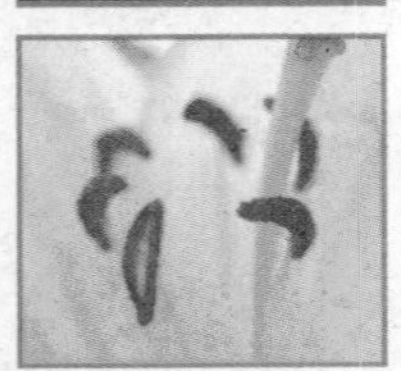

pollen powder that helps make new plants

protection keeps things safe from harm

R

reflect give back an image

reptile an air-breathing animal that has dry skin covered with scales

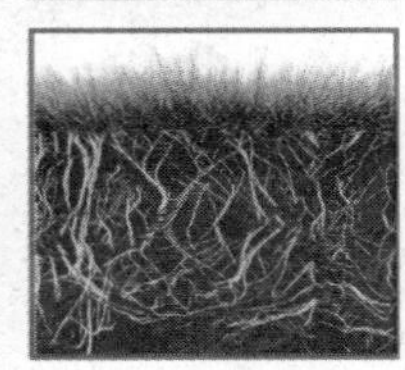

root a plant part that keeps the plant in the ground, stores food, and absorbs water and nutrients

S

season one of the four parts of the year with different weather patterns

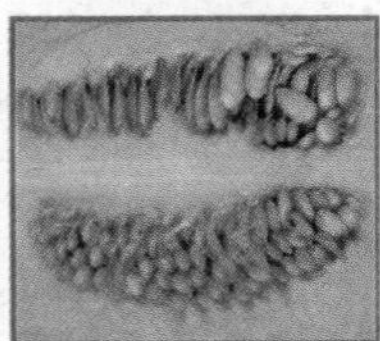

seed a part of a plant that can grow into a new plant

seedling a young plant

shadow a dark shape that is made when a source of light is blocked

signal a sound or movement that gives information

sound a form of energy that comes from objects that vibrate

spring the season after winter

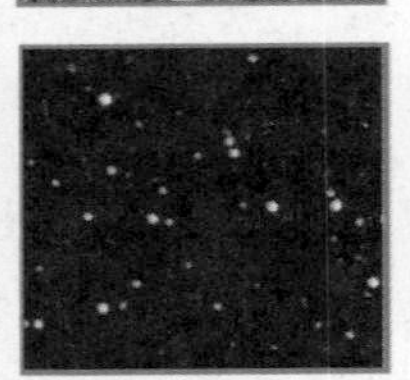

star an object in the sky that makes its own light

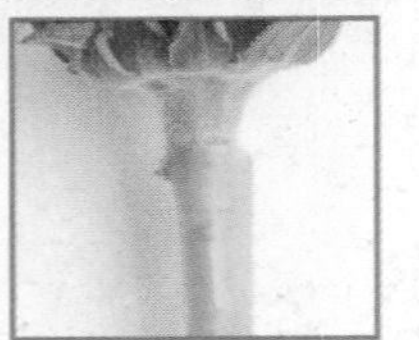

stem plant part that holds up the plant

structure a part of something

summer the season after spring

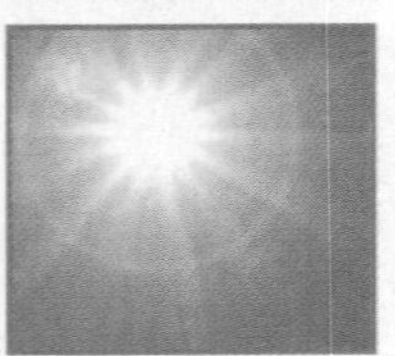

Sun the star closest to Earth

sunrise the time of day when the Sun rises above the horizon

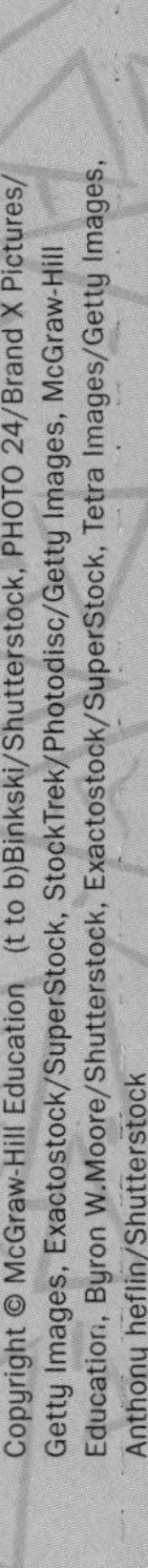

sunset the time of day when the Sun descends below the horizon

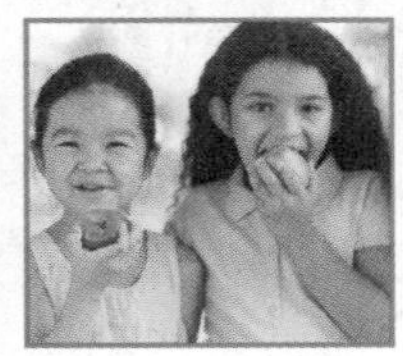

survive to live and grow

T

trait a feature of a living thing

translucent when some light can pass through

transparent when light can pass through

V

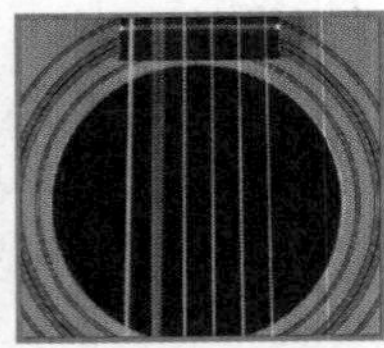

vibrate to move back and forth

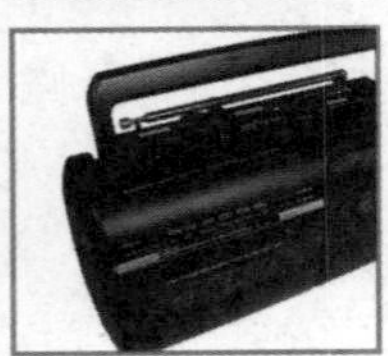

volume how loud or soft a sound is

W

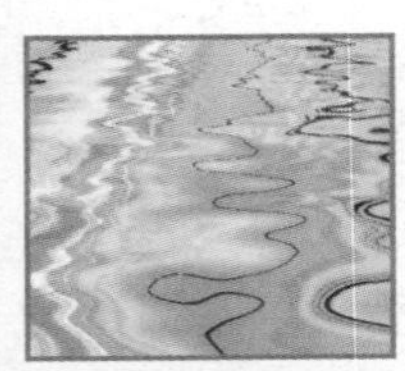

wave a movement up and down or back and forth

winter the season after fall

Index

cut on all dashed lines

fold on all solid lines

leaf

A ______ is the plant part that uses sunlight and air to make food.

Plants use their ______ to make food.

A ______ is the plant part that makes seeds.

A ______ is the plant part that makes food from sunlight.

The ______ is the plant part that holds up the plant.

The ______ is the plant part that keeps the plant in the ground.

cut on all dashed lines

fold on all solid lines

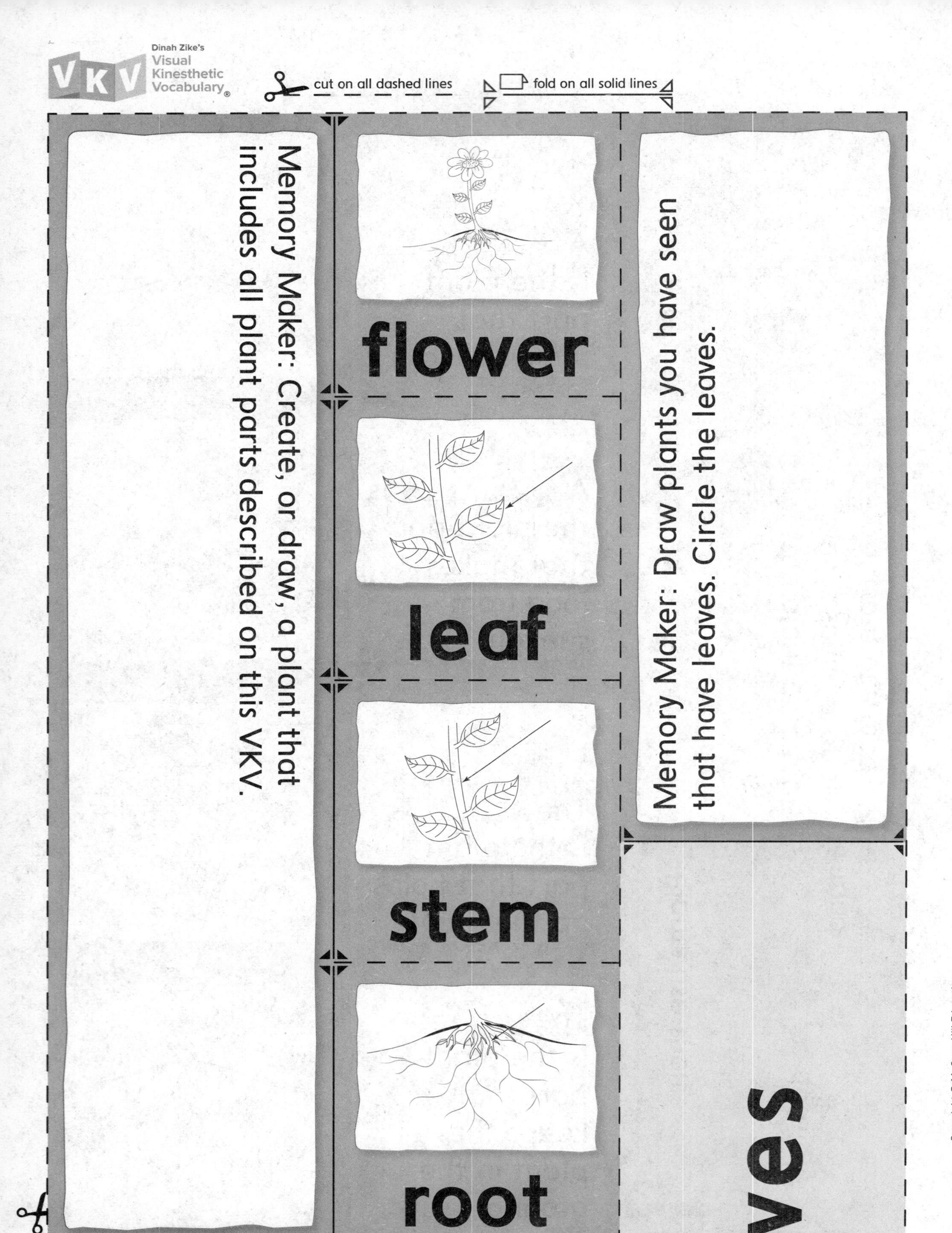

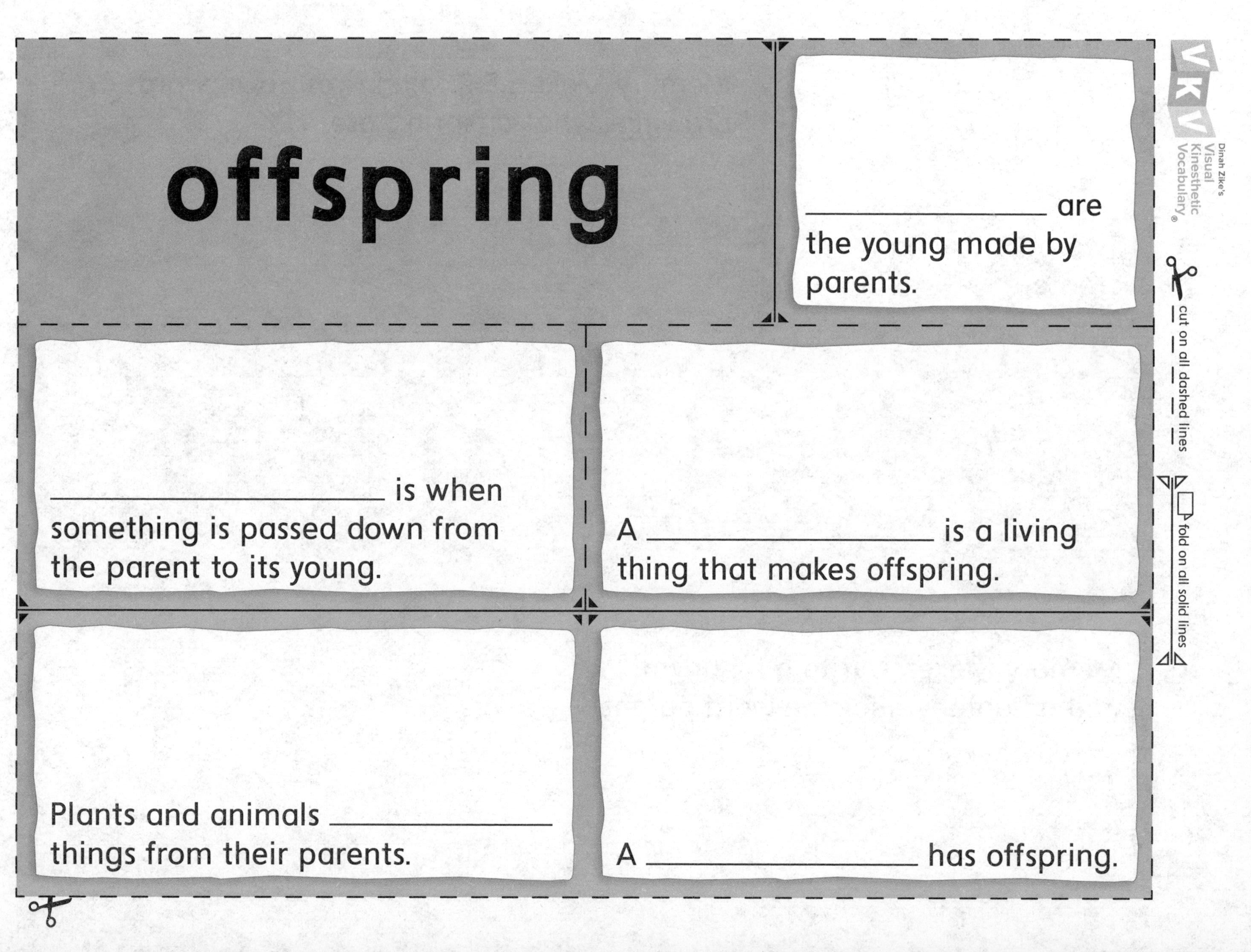

offspring
______ are the young made by parents.
______ is when something is passed down from the parent to its young.
A ______ is a living thing that makes offspring.
Plants and animals ______ things from their parents.
A ______ has offspring.
Dinah Zike's Visual Kinesthetic Vocabulary®
cut on all dashed lines
fold on all solid lines

Dinah Zike's Visual Kinesthetic Vocabulary®

cut on all dashed lines

fold on all solid lines

Two parents can create ________.

Memory Maker: Explain in your own words or drawings what offspring are.

parent

inherit

Memory Maker: Write a synonym and an antonym for the word parent.